# The nature of Nature

German Navarro

# The nature of Nature

# DEDICATION

Writing about Nature and being a
Dutchman, I have to dedicate this book

*To them who die while they enjoy,*
*to them who love while they care not,*
*to them who weep while they laugh,*
*to them who endure while they exalt.*

*To those poor, little ones,*
*who don't get born and yes, who die,*
*to them who they dispose of,*
*For being ill, for being old.*

# CONTENTS

# ACKNOWLEDGMENTS

Not having in mind any particular human being to acknowledge for the writing of this book, I acknowledge Nature for its absolute refusal to get determined by human mind.

# Preface

This book deals with numbers, particularly with the number zero. This number has no doubt played an important role in the development of modern science. Science however has a double meaning in this book: engineering on the one hand, and the search for fundamental variables of Nature on the other. The former is considered to be applied science, the latter is viewed as man's effort to explain the great mysteries of the universe we live in. The author's position is that the number zero "behaves" differently in engineering and in fundamental research.

While dealing with the scales of measurement, special attention is given to the number zero in view of the absolute importance this number has in measurements at ratio level. What would happen to the fundamental research of Nature should we ever get dispossessed of the number zero? In the pages that follow, the author tries to give an answer to this fundamental question.

The reader will also come across a new approach to prime numbers. After primes have been redefined, this new approach leads to a new solution, a solution that disregards completely the prime numbers' indivisibility. And since computers are the only instruments capable of finding prime numbers in an efficient way, an algorithm is offered, that the author qualifies as the fastest possible. Whether or not that is the case, it is up to others to judge. The prime numbers are found, hidden from the human eye in a numerical sequence that is completely unknown in mathematics. The primes, it appears, occupy predictable places in that obscure numerical sequence.

The author has tried to make the content of this book accessible to everybody and therefore the use of mathematical jargon has been avoided as much as possible. Nonetheless the reader that avoids numbers by instinct will come across one or two passages that will require some effort.

Finally, about the title of this book: for good reason we have decided that the only way to get to the essence of Nature is through the use of numbers. No numbers, no knowledge of Nature. This book, being actually a fundamental reflection on the essence of the port that gives us access to Nature (the number zero) could bear no other title than "The nature of Nature".

# ZERO

In *The Lost Civilizations of the Stone Age*, Richard Rudgley describes masterfully the way human kind learned how to count. The Middle Eastern tribes whose humble way of life he describes were not in search of the great mysteries of Nature, but were rather tentatively trying to ascertain each individual's share of the commonly possessed livestock.

Those tribes had changed their way of life profoundly. They were no longer nomads, wandering from one water source to the other, hiding from predators, searching for new fertile land to pasture their commonly possessed stock. Instead, they had chosen to settle down in the vicinity of water, both from heaven and earth, water that they desperately needed were they to survive. In the new settlements, the individual human began to count, networks were established, family lines were neatly drawn. The new communities gave birth to "the individual right of property", something unknown while humans were still nomads. They began to concern themselves with mastering the way to express each individual's or family's wealth in terms of the amount of objects or stock they possessed, and numbers did just that.

Before they learned how to count properly, it is probable that our ancestors had a rudimentary counting system based on intuition. We, modern humans, possess a nature-given, visual skill to count. Look at the sky and observe the birds flying: if it is one, you know it is only one. If it is two, you know it. If it is three, you don't have to count them. If it is four, you'll have to make sure it is four. By about five or six, the problems begin. You have to count them; otherwise you do not know how many birds you saw. Apparently, our ability to count intuitively has a threshold, and quite a low one for that matter.

So, knowing for certain that the human beings

studied by Rudgley were absolutely comparable to modern man, there is no reason to believe that their ability to count intuitively differed from ours. It is imaginable therefore that they had distinct words to indicate the amounts of objects that they could discern quantitatively from one another in an intuitive way. If we agree to put this limit at a conservative 6, then it is possible that they had the ability to count from 1 to 6 but that anything equal to or larger than 7 could only have been spoken of as "much" or "several" or "many".

However it might have been, once the former nomadic tribes became sedentary, culture began to evolve and with culture knowledge and with knowledge economics and with economics numbers. It is superfluous to repeat a story so magnificently told by Rudgley. Instead, a schematic view of what happened to numbers will be presented here: the new civilizations did indeed develop a sophisticated counting system, so sophisticated that it only marginally differs from ours, thousands of years later. Numbers were born.

The individual Stone Age man possessed sheep, goats, donkeys, all sorts of other domesticated animals and plants in quantities diverse from those of his neighbours, and so arose the absolute necessity to distinguish 17 goats both from 16 and from 18, as intuitive, eye-counting did not suffice any more to express and differentiate their possessions.

The counting system developed by the Middle Eastern tribes was meant to express *discrete* quantities. The lowest number they knew was therefore 1; zero did not belong to their counting system as zero is the absence of any quantity and the counting system was aimed at expressing quantities, not the absence thereof. Shortly, counting was nothing other than applying numbers directly to the items that were counted almost in a predicative way. "The number of sheep I possess is (equal to) nine", "the number of children my brother has is (equal to) three". The items that were counted were discrete, not divisible: one quarter of a goat or two thirds of an arrow were still waiting to be invented.

We, modern humans have a tendency to think easily about numbers. By the time we reach the age of ten or twelve years, most of us are capable of counting indefinitely, not realising that it could have taken the human race one thousand years to learn how to count from one to one thousand, so to speak. Apparently, our brain was made to deal with numbers, certainly not to invent them.

Another thing we don't realise is the huge difference that exists between *counting* and *measuring*. Although strictly related with one another, they are two different activities of the human mind. Whereas counting, as it has just been stated, applies numbers directly to bare objects, measuring is based on an intermediate social agreement. Measuring presupposes that a society, large or small, has reached agreement on a standard that is to be taken as the unit of measurement. Again, taking the example of the ten or twelve year old child: show him two coins, one of 1 Euro and another of 2. Ask him how much is that and he will give you the right answer with no hesitation: three Euros. What the child does not realise is that he has done two separate mental operations: first of all, he has *counted* the coins, being two, and in the second place he has *measured* each coin's value being one and two Euros, yielding three Euros as the right answer. What has happened in this instance is that the child is aware of the social agreement that has assigned a particular value to a particular piece of metal coined in a particular way, accepts that agreement, evaluates (= measures) each coin's value and counts them.

Counting coins and counting values: they both go: one, two, three … They seem to be identical, but in reality they belong to two separate worlds. If it took the human race one thousand years to learn how to count from one to one thousand, as earlier suggested, another thousand years were needed to socially accept a vessel of some sort as being the standard of measurement for liquids and grains (again, so to speak).

It is probable that the first standards upon which social agreement was ever reached expressed a *small amount* of "something". That "something" was probably a commodity that had either liquid or granular properties and that therefore had to be contained in a bag or vessel of some sort in order for it to be transported or indeed bartered or even sold.

That such was indeed the case is commonly agreed upon in history. Mesopotamians, Greek, Romans ... they all knew all sorts of standards accepted by society and ratified by law.

What happens mathematically when a society goes over from counting to measuring is that counting no longer is applied to *concrete objects*, indivisible entities that are essentially identical to one another and interchangeable, but rather on some *abstract measurement unit* that represents or contains a standard amount, quantity, weight, length ... that socially has been agreed upon. Society goes over from one, two, three *houses* to one, two, three *feet*. Counting and measuring, they both operate on numbers but yield different answers: after counting, the resulting number gives a particular amount of individual objects whereas after measuring, the result expresses the number of measurement units (or the fraction thereof) that such an amount contains or represents. The *counting system*, earlier applied directly on bare objects, becomes a *measurement system* in that it is a *measure* that has acquired an intermediate role between matter and numbers. Measuring is counting abstract objects, so to speak.

We don't know how exactly fractions did come into existence but it is very well imaginable that fractioning was a consequence of measuring. As it was said, it is probable that the first measurement standards were intended to express or contain small quantities of the substance or commodity being measured. As human settlements grew ever larger and larger, there probably arose the necessity of constructing greater "containers" that were nothing other than a multiple of the smaller ones, the first units of measurement upon which social agreement had been reached. From there to a reverse situation, where the larger unit of measurement becomes the standard and the smaller, original one becomes a "fraction" thereof, a very small step is needed.

During thousands of years numbers were used only for *counting* purposes, but by the time the Mesopotamian culture arose, we know that numbers were also used as an instrument of *measurement*. The step from counting to measuring is a huge one: for the former all one needs is a discrete amount of units to be counted, whereas for the latter social accord is needed: every Roman knew (by approximation) what a *pes* or a *congius* was. The social agreement that is needed to use numbers as an instrument of measurement creates the necessity of the use of fractions: Romans, for instance, knew both small and large units of measurement. Using the smallest units as a departure point, they expressed the larger ones as multiples thereof, whereas departing from the larger units they expressed the smaller ones as a fraction of them.

Speaking in today's terms, while human beings were able only to count they were able to express or manipulate *discrete variables,* whereas with measurement, *continuous variables* appeared, continuity being nothing other than the ability of a variable to acquire not only whole, but also fractional values. Actually, what *measuring* did to *counting* was that the former created a whole range of new (fractional) numbers between any two consecutive digits used in the old counting system but most important of all: measuring opened the infinitely deep gap below 1, the lowest of all numbers that the counting system knew. Modern mathematicians would undoubtedly say that the gap between 0 and 1 was finally covered, an assertion that is only marginally true.

Going back to counting, it is hardly imaginable that its inventors realised how powerful this instrument would become in history. In due time the numbers themselves were to become the object of research. And even as far as these pages are concerned, they will deal with some aspects of our ancestors' counting system that have survived time as well as with some relatively new, questionable additions.

Counting up, to the right side of the number one as it were, mathematicians soon concluded not only that they were unable to find an upper limit to counting but that indeed the counting system as such had no upper limit at all. The greatest possible number could never be found: should anyone claim to have found it, all one had to do was to add 1 to it, proving the claim to be false.

Counting down was quite another matter and it all has to do with number zero. Modern mathematicians are inclined to take zero as the departure point of the counting system, not only to its "right" but also to its "left". Numbers on its right are positive, numbers to its left are negative. Should one try to find the least possible number, one can apply the same reasoning used for positive numbers: you can count down as long as you want, a limit will never be reached.

The problem with zero in modern mathematics was its unquestioned entrance into the kingdom of numbers. What is more, zero was placed at the very centre of numerical creation surrounded by an infinity of numbers on its right side as well as on its left side. Zero was actually granted the role of "king of numbers", a title that zero certainly does not deserve.

The most fundamental question about zero is: is zero really a number? Obviously, in a *discrete* counting system there is no place for it. Saying, as mathematicians tend to do, that a person that has no sheep, actually possesses zero sheep is an aberration to say the least. As it was earlier stated, the lowest possible number in a discrete counting system is 1. Such a system knows no fractions on the one hand, and on the other, a discrete counting system is meant to count discrete quantities of entities, entities that are indivisible. This all means that, concretely speaking, there is no such thing as a fraction of a sheep or indeed a nonexistent or "empty" sheep to which "number" zero can be attached to in order to start counting from there. The zero sheep exists only in the mathematician's mind.

Parallel to the question of the existence of zero as a member of the *"counting"* family of numbers runs the question of its existence as a member of the *"measuring"* family. It was earlier stated that measuring introduced fractioning and with fractioning an attempt was made to bridge the infinite gap between one and zero. As it has just been said, mathematicians accepted zero as the "natural" number placed half way between positive and negative amounts. If zero has been rejected as a member of the counting family, then there arises the question of the legitimacy of its existence within the measuring family of numbers. Well, if it is unacceptable to speak of zero sheep, is it legitimate to speak of zero feet, zero pounds, or zero Euros?

At the very face of it, there is nothing wrong with giving a positive answer to these questions. Actually, the tremendous advance of science wouldn't have been possible without the presence of zero in our measurement system. For engineering purposes we can very well pretend to be able to depart from zero speed, zero weight, zero time. And as long as we keep measuring our small world, building instruments, observing the behaviour of matter on Earth and its surroundings, there is nothing wrong with it. The problem arises when "science" jumps from engineering to rampant speculation: big bangs, iterative universes, speed limiting ... all these thoughts can't be conceived without zeroing one or more of the fundamental variables that constitute Nature, be it time, be it energy, be it mass ...

But is it valid to speak of zero time, zero energy, zero light, or zero mass? Mathematicians have absolutely no problem with these questions; neither do they have a problem with their answers. Yes, they say, it is utterly legitimate to speak of zero time, zero energy, zero mass and so on. But then, what do mathematicians mean when they zero these fundamental variables of Nature? Obviously, while attributing zero to energy, they refer to a situation in which the "amount" of energy indicated is nonexistent, which is inherent to the very definition of zero. There is nothing "there". The only thing that is "there" is emptiness. And, when they speak of zero time there is nothing "there". The only thing that is "there" is also emptiness.

This concept of "emptiness" is inherent to the very essence of measuring. While measuring some particular variable, the human mind concentrates solely on the variable being measured, pretending as it were that the rest of reality does not exist. When we speak of 20 kilograms of salt for instance we implicitly abstract from measuring any other variable and indeed, we even abstract from the presence of any other variable. So, while measuring a particular amount of salt and concluding that this amount is equal to zero we actually conclude that "emptiness" is there. There is nothing "there". Not only no salt, but also no light for we did not intend to measure any light; no mass, for we did not intend to measure any mass; no energy, for we did not intend to measure any energy ... so, measuring energy and finding a zero amount of it yields the same result as measuring mass and finding a zero amount of it or measuring salt and finding a zero amount of it. Generally speaking a zero quantity of any variable is identical to a zero value of any other variable: zero time is perfectly identical to zero energy, to zero mass, to zero distance.

So, zeroing any fundamental variable of Nature leads to the same result. And this result is emptiness, as zero is emptiness, the negation of everything. Departing from energy for example, zero energy means also zero mass, zero time, zero light, zero gravity ...Zero cannot be "applied" to any variable in Nature without "applying" it to all other variables that constitute the same Nature. This is due to the overwhelming, "destructive" power of zero. Once Nature is "there", even with the tiniest possible "amount" of anything, then and only then has zero been left behind, emptiness is no longer "there". That tiny "amount" of energy, or time, or mass ... lends itself then to be measured, reality is back, and Nature is again present.

After these considerations it becomes clear that choosing zero as a departing point for any abstract, non-engineering related measurement means:

1. Choosing as departure point a "place" outside the universe as it is known to us: a "place" with zero energy, zero light, zero mass, zero gravity ... How *on earth* mathematicians are then able to "jump" from there to this universe, at least to this earth of ours, full of energy, full of light ... remains as yet an unexplained miracle.

2. Since zero energy is the absolute absence not only of energy but also of anything else, it is obvious that zero energy is absolutely equal to zero gravity, to zero light, and so forth, for those "zero states" also represent the

absolute absence of everything. Well, taking a step, as mathematicians theoretically do, from zero energy to the smallest possible amount of energy is tantamount to "jumping" from "zero light" to the smallest possible amount of energy, from "zero gravity" to the smallest possible amount of energy, from "zero space" to the smallest possible amount of mass … in a few words, a "jump" from zero to Nature results in a continuity between energy and light, between energy and time, between space and mass … obviously, an aberration, to say the least.

Finally, there is a fundamental question regarding the presence of zero within the *measurement* system. As it was earlier stated, measurement made fractioning possible and fractioning did open the way to explore quantities below the number 1, the absolute lowest limit admitted by the *counting* system.

It is well known, and admitted by everybody, that our counting *and* our measuring systems do not know an upper limit, meaning that either the human mind is incapable of finding the greatest possible number or that such a number does not exist. In either case it is far beyond our mental capabilities to get a hold of it.

The question then arises: what about the smallest possible number? Well, once zero got introduced (or rather: smuggled)  into our numerical systems, mathematicians simply "flew" over zero and declared the problem of the smallest possible number to be simply the mirror image of the greatest possible one. In other words, the smallest possible number is equal to the greatest possible one with a negative sign attached to it.

In reality, once fractions come into existence the smallest possible number has to be sought below 1, as near to zero as possible. Once fractions have entered the human mind, one cannot disregard the presence of zero as being the absolute absence of any quantity at all. Well, no matter how deep you dig underneath 1, you will never ever be able to lay your hand on the smallest possible number for you will never ever get to zero. That is where the smallest possible number "resides" and not at the very end of the negative numbers as mathematicians state.

From what has just been said one is bound to conclude that zero is as disqualified to enter the family of *measuring* numbers as it was to enter the family of *counting* numbers. In a few words, zero is a good thing to have … for engineering purposes where its usefulness has been proven. In the utterly prestigious world of the mathematicians, however, zero is a good thing … to be done with.

# SCALES OF MEASUREMENT

When modern thought began to evolve in Europe in the 15th and 16th centuries, it was quite natural that astronomers, physicists, and mathematicians, began to create a jargon of their own that allowed them to communicate with one another. As time passed by and each discipline specialized in creating their own concepts and way of thinking and researching, they developed a common meta-language as it were and that was mathematics, the realm of numbers. Mathematics allowed them to understand one another not only alone from within, but also between disciplines.

Science as a whole became the queen of

numbers and went much further. Scientists not only got hold of all concepts related to astronomical, to physical, and to biological research, but they also claimed for themselves two fundamental, exclusive rights:

- The right to create instruments with which to enhance their research.

- The right to deal properly with numbers.

Towards 1900 though, there emerged a new breed of researchers that "pretended" to be able to deal with numbers too. As the body of social sciences begun to emerge and as no one could deny them the right to exist, there arose the discussion of the measurability of certain variables, human and otherwise, that one could not measure or observe by the use of a stick, a balance or a microscope; i.e. variables that were not measurable by the same means scientists were using. Social researchers had observed for instance the great differences that existed concerning the degree of intelligence amongst human beings but found themselves empty handed when it came to measuring these differences. There were no available instruments to measure them.

The new scientists felt they had no one's permission, let alone the mathematicians', first to develop a whole set of instruments that allowed them to perform their own measurements and research, and secondly to deal with numbers in their own way. In a few words, they claimed the right to found a new body of scientific knowledge alongside the existing one, that of the old science. The old scientists, not having any means to inhibit the creation of the new instruments of measurement, resented very much the use of numbers by the new breed of researchers. The realm of numbers belonged to the old science and the newcomers had better keep away from them.

As it was explained in the first chapter, the historic step that was needed to go from *counting* to *measuring* was crucial to the awareness of the existence of continuous variables, of fractions and in the end, of the number zero as it was interpreted by science. When the social sciences appeared, counting and measuring began to belong to the public domain, and the number zero as well as discrete and continuous variables were all known.

After a long struggle between the old and the new science there emerged an agreement about who should be allowed to use what, while dealing with numbers. The final decision centred firstly on the intensity to which variables allowed themselves to be measured, this intensity being nothing other than their ability to acquire fractional values between (any) two consecutive integers, and secondly on their ability to acquire the value of zero.

The reached agreement determined that there were 4 different scales of measurement:

- The nominal scale.

- The ordinal scale.

- The interval scale.

- The ratio scale.

Although these four scales of measurement belong to public knowledge, a short description of them shall be given here. Before entering into the matter, one should not forget that in all four cases, we are dealing with *measurements* and that all measurements, as has been stated before, have to do with numbers: measuring is equivalent to attaching numbers to the values that a variable acquires.

First of all, the *nominal* scale: the numbers attached to the values that a variable measured at this level can acquire are actually not numbers at all. Some marketer, for instance, researches the public's preferences for green, red or yellow apples. It is a research based solely on colour. After "measuring" the preferences of thousands of people he is ready to analyse his data. As computers have less trouble dealing with numbers than with words, the researcher decides to "attach" the number 1 to "green", the number 2 to "red" and the number 3 to "yellow". As measuring is nothing other than attaching numbers to the different values that a variable can acquire, we can say that in this case a true measurement has taken place: the numbers 1, 2 and 3 have been attached to the three different values that the variable "colour of an apple" can acquire (supposing that there are only three of them), green, red and yellow.

It goes without saying that the numbers used at this level are not used as such. They are there only for convenience. Attributing numbers 1, 2 and 3 to the different colours that an apple can have has nothing to do with attributing quantitative properties to the colours they are attached to. As a matter of fact, the three numbers can be used interchangeably and indeed, each one of them could be replaced by any other number with no effect whatsoever on the final results. No wonder then that performing measurements at this level has been called *nominal*. Numbers are present only nominally, not really: they represent no numerical values at all.

Measuring at the *ordinal* level is a little more complex, but as the name itself indicates, measuring at this level has to do with *ranking*. We all learned at grammar school that numbers are to be divided into cardinal and ordinal, the former indicating quantities, the latter indicating ranking. So, *two* is a cardinal whereas *second* is an ordinal number. An example will clarify things.

A psychologist researching the development of the visual acuity of human beings decides to take measurements on 4 groups of 10 individuals each:

- 10 infants aged 1 to 3 years.

- 10 adolescents with an average of 12 years.

- 10 adults, all of the age of 49 years.

- 10 elderly with an average age of 67 years.

There are several things that are worthwhile noticing. First of all, it lies in the nature of things to rank the four groups the way we have just done, with the infants first and the elderly last. The four groups have been ranked according to age and if we decide to order them increasingly, from lowest to highest, then it becomes "logical" to attach to any group a rank higher than the previous one but lower that the next. Second, and this is extremely important, one should notice that the distances between the age groups, expressed in years, are UNequal to one another: the "distance" between groups 1 and 2 is equal to 10 years, that between groups 2 and 3 amounts to 37 years, whereas groups 3 and 4 are separated from one another by a "distance" of 18 years.

The measurement that takes place at this ordinal level consists simply of "attaching" ordinal numbers to the different values of the variable (remember: measuring is nothing other than attaching different numbers to different values of a variable) that one decides to research with one condition though: that the values that are taken are NOT equidistant from one another. In other words, at this *ordinal* level of measurement there should be some sort of chaos, although not absolute chaos.

The third measurement scale occurs at *interval* level. The difference between measuring at ordinal and at interval level is actually very simple: At the interval level of measurement the measurement points on the researched variable are taken at equidistant points from one another. Repeating, for the sake of simplicity, the last given example, the said psychologist redirects his research to these four groups:

- 10 infants all aged 5 years,

- 10 adolescents with an age of 15 years,

- 10 young people aged 25 years,

   - 10 persons all of the age of 35 years.

The most important thing to know about this example is that the measurement points on the variable "age of human subjects" have been taken equidistantly from one another: they all differ by 10 years.

Finally, measurements at the *ratio* scale. This theory says: these are measurements taken on variables that admit equidistance as well as a value of zero. Trying to explain this somehow mysterious definition of the ratio measurement scale, let us go back to the classical example of measurements at *interval* level. It has been said time and again that temperature can at most be measured at interval level for these reasons: although temperature lets itself be measured equidistantly (the "distance" between 9 and 10 degrees Celsius is said to be equal to that between 10 and 11 degrees), the variable called "temperature" cannot achieve a value of zero. This point of view has been maintained even after science has found out that a zero temperature does exist (-273.15 degrees on the Celsius scale). For some reason though, science does not allow us to say that a temperature that is 2 degrees higher than -273.15 is twice as "warm" as a temperature situated only 1 degree away from the absolute zero point.

Turning back to the *ratio* scale of measurement: in order for a variable to be measured at ratio level it has to admit equidistance as well as a zero value of measurement. Length, for instance, states science, is a variable that lends itself perfectly to be measured at ratio level: one can very well think of a body of zero length, says science, but once one begins measuring length at the "right" side of zero, measuring the length of a body obeys the same laws that govern the measuring of continuous variables, as has been explained in chapter 1: a variable, in this case length, is allowed to take not only integer values but also all fractional values between any two consecutive integers.

The most vulnerable aspect of the *ratio* scale of measurement is of course its need of a zero measurement point. As it was stated in chapter 1, while "acquiring" a zero value the variable that acquires it loses its own identity. Zero light means not only no light at all, but also nothing at all: emptiness. Zero gravity means not only no gravity at all but also nothing at all: emptiness as well.

Taken as a whole, the entire discussion regarding the 4 scales of measurement was made mainly with one purpose: to distinguish old from new science. "True" science, we were told, operates only at *ratio* level; whereas the new "human" sciences were granted a free way to operate at *nominal*, *ordinal* and *interval* scale, the big difference being that ratio variables can acquire a zero-value, while ALL other variables do not, or so "old" science told us. It is however the same "old" science that, be it accidentally or not, one given day discovers that one of the most fundamental variables of our universe, temperature, reaches a point beyond which it can't decrease any further. It is an absolute zero point, one would say, determined by Nature itself. And still, says science, this nature-given absolute zero point cannot be treated as a zero point of measurement simply because it does not fit into our numerical system: a temperature of absolute 2 degrees does not represent a temperature that is twice as "warm" as a temperature of absolute 1 degree. Temperature does increase from absolute 1 to absolute 2 degrees, says science, but it does not double. The question then arises: at which point is there double a temperature of absolute 1 degree? At absolute 3 degrees? At absolute 45 degrees? At absolute one million degrees? Science does not know.

The paradox about the scales of measurement is that, normally speaking, Nature does not give us free access to its own absolute zero values. This seems to occur, though, with temperature and all we have to say is: we can't handle this zero, so temperature is NOT a ratio variable, it can be measured only at interval level. On the contrary, there exist scores of variables whose absolute zero value Nature keeps hidden from us and we nevertheless pretend to be able to measure them at ratio level. Take length, which was mentioned earlier. We feel free to assign to this variable a zero-value whenever and wherever we deem it necessary. No blinking whatsoever. Besides, departing from this zero-value, we simply take a stick, lay it on the ground once, twice, three times, draw marks on the sand and proudly announce: the second mark doubles the first mark's length, the third triples it etc. etc. etc. What "on earth" makes the variable "length" that special that it allows us to manipulate and measure it any way we wish? What has "length" that "temperature" misses? Has it something to do with the physical shape of the instrument we use to perform measurements? Has it to do with the nature of the variable itself? Has it to do with the human senses through which "knowledge" of a certain variable enters into our brain? Has it to do with the nature of our numerical system? Why is a distance of two feet twice as long as a distance of one foot, but a temperature of two degrees (departing from absolute zero temperature) is not twice as warm as a temperature of 1 degree?

After all these considerations there arises a final question: suppose one day Nature decides to disclose to us the absolute zero point values of all the fundamental variables that make up our universe: zero space, zero time, zero energy ... Take then time. Set our best atomic clock running at 00:00:00. Is 00:00:02 twice as long as 00:00:01? If it is, it means that the abysmal jump from no time to time, something that had to occur between

00:00:00 and 00:00:01 had to repeat itself somewhere between 00:00:01 and 00:00:02, between 00:00:02 and 00:00:03 ... so at each new measurement point, time has to jump from nonexistence into existence; otherwise things do not add up. And if time jumps into existence each time a new measurement is done, it means that time somehow had ceased to exist prior to that; otherwise, it could not jump into existence time and again. If instead of taking time, we take any other fundamental variable of our universe, be it energy, be it light, be it speed ... we are confronted with the same basic problem: once any fundamental variable of our universe comes into existence by leaving "behind" its zero measurement point, it has to come into existence time and time again. Is this what really happens in Nature? It must if we stubbornly keep alive our conviction that our numeric system is the best instrument we have at hand to interpret Nature and that a variable that really deserves this name should be able to acquire a zero measurement point.

Admitting, or rather requiring, as we do, that the fundamental variables that constitute our universe have a zero measurement point changes dramatically our conception of reality. We are then left with an intermittent universe that comes into existence time and time again at a rate that is only dictated by the speed at which the universe lends itself to be measured. And this is where the concept of measurement becomes central.

Who is in charge of performing this measurement? In the present context there is but one answer to this question: the human mind. WE seem to be able to decide which variables are continuous and which are not, WE seem to be able to decide which variables must have a zero measurement point and which do not, WE are the only beings capable of making any measurement at all, so it is up to US to provide an answer as to how often either WE decide to measure the fundamental variables of Nature, or how often Nature lends itself to be measured by us.

What has just been said is what we shall call here the phenomenon of *intermittency*: while measuring the fundamental variables of Nature, admitting that each one of them possesses a zero measurement point, on our way from zero towards measurement point 1, we necessarily have to cover the abysmal gap between zero and the tiniest possible "amount" of the variable concerned. Measuring forth, we will eventually get to measurement point 2. Well, in order to arrive at 2, the "distance" covered between 1 and 2 should be equal to the "distance" between 0 and 1; for if they are unequal, we have not arrived at 2 but at some other measurement point. This means that the "distance" between 1 and 2 should also cover the abysmal gap between a non existing and an existing variable, just as it did between 0 and 1. Well, if "leaving" 1 towards 2 means departing again from zero, this necessarily means that by arriving at 1, the variable's value automatically returned to zero, the measurement point where the variable that is being measured is non existent. So, while measuring, every time a measurement increases by a value equal to unity, the variable's value has to come back to zero, returning back into existence immediately thereafter. The rate of this periodicity, or *intermittency*, is determined by the variable's "amount" that we have decided to define as the variable's unity.

The absurdity of *intermittency* denies us the right to perform ratio level measurements on the fundamental variables of Nature. This compels us to conclude that we lack the basic intellectual skills that enable us to know the zero values of the fundamental variables of Nature, if indeed these points exist or ever existed. Any time we state, in the name of the human race or otherwise, to be able to "jump" over the zero value of any fundamental variable of Nature, we are behaving like gods: beings that pretend to know everything, even the "unknowable".

The general conclusion of this second chapter is not essentially different from the earlier one, taken at the end of chapter 1. If it was stated there that zero was a *fremdes Körper* in the world of counting and measuring, here one is bound to conclude that accepting zero as a measuring point of the fundamental variables of Nature, taken in their fundamental context, leads to a bizarre universe that intermittently jumps from no existence into existence and then back into nonexistence again, this last step having the sole aim of meeting our absolute requirement that 1 plus 1 must be equal to 2, under any circumstance.

One of the ways to avoid this continuous back and forth jumping of Nature from nonexistence into existence and then back again into being could be to question the fundamental ability of our numerical system to deal properly with Nature. After all, this elementary 1, 2, 3-sequence, a humble stone-age invention, was made essentially to count goats, not the speed of light immediately after its release from absolute darkness, or the enormous amounts of energy released after two black holes have collided with one another at the other side of the universe. One goat and another goat taken together are undoubtedly two goats. Absolutely nothing stands in the way to add them up as they are totally similar. But, going back to the situation where light was two seconds old, was its second second of existence totally equal to its first second? Absolutely not. So when light was two seconds old, light was not two seconds old unless light somehow disappeared after the first second and it miraculously jumped back into existence at the beginning of the second second. This again, takes us to the intermittent universe discussed earlier. If we are prepared to accept this, then it is true that when light was two seconds old, light was indeed two seconds old. Chapter 3 will hopefully throw some new light into this little black hole of ours called the numerical system.

We began this chapter with the aim of explaining the four levels of measurement that science and human sciences have agreed upon. Science has claimed for itself the exclusive right to perform measurements on variables that admit a zero measurement point. Any variable that can be measured at such a level is "theirs". Now we are ready to conclude that this is true only while we remain in the world that belongs to the engineers. If we transcend that little world and come into the realm of the fundamental variables of Nature, measurements by human beings at the ratio level are utterly impossible.

# PRIME NUMBERS

Before getting into the matter, a general observation should be made. Although these pages are also intended for the mathematician, they are not written in the concise language mathematicians tend to use to express themselves. As it happens, most people abhor numbers and they seem to have a tendency to put aside whatever it is that contains them. On the other hand, the concepts explained in these pages are self-explanatory, so there is no need of a sophisticated jargon to clarify them.

If the most important number to science seems to be zero, to those outside science the most important of all numbers is by far number 1. This little number, the smallest of all integers, has two properties that make it indispensable in our counting system:

- To begin with, 1 is the initiator, the door to our counting system.

- Then, and most importantly, 1 is unity.

In its role of unity, the number 1 is actually the engine that propels our counting system limitlessly onwards: every single number after 1 is equal to its predecessor plus 1:

- $1 + 1 = 2$

- $2 + 1 = 3$

- $3 + 1 = 4$            **SEQUENCE 1**

- $4 + 1 = 5$

- ...

Because the number 1 possesses such a property, it has also been called *unity*, derived from the word *unus*, Latin for "one". This is the way science looks at the number 1. Everything is neatly in balance.

There is another quality of this tiny little number that makes it extremely interesting. The following sequence will make it clear:

- $(1) + 1 = 2$

- $(1 + 1)$ **+ 1** = 3

- $(1 + 1 + 1)$ **+ 1** = 4          **<u>SEQUENCE 2</u>**

- $(1 + 1 + 1 + 1)$ **+ 1** = 5

- ...

Both sequences are mathematically identical even though the first digit in the first sequence has been replaced in the second sequence by the number of "ones" that it contains. So "2" for instance in sequence 1 has been replaced by "$(1 + 1)$" in sequence 2.

For the purpose of this discussion, the first digits in <u>sequence 1</u> and the parenthesised "ones" in <u>sequence 2</u> are called the *base*. The digit at the right side of the "=" sign in both sequences will be called the *target*. To take another example, "(1+1+1+1)" and "4" are called the *base* and "5" is the *target*.

As it has just been said, science prefers sequence 1 to explain the numerical system: every digit in that system is equal to its predecessor plus unity. On the other hand, if one decides to look at sequence 2 in a particular manner, things seem to get out of balance. Consider the number 2 as the target in <u>sequence 2</u>. It takes "a whole 1", i.e. 100% of its predecessor to get from 1 to 2. Counting further, from 2 to 3, it takes "only" 50% of 2 to get to 3, it takes 0.33% of 3 to get to 4 ...

Generally speaking, if one decides to explain the stepwise increment of our counting system in terms of the percentage of the *base* that is needed to get at the *target*, our counting system is unbalanced and very much so as that increment becomes increasingly smaller. In a few words, science states "it takes unity to get from *base* to *target*" and things are in balance; whereas an alternative explanation says "it takes a decreasing share of base to get to target" and things are out of balance.

We tend to take it for granted that the number 1 is the only possible unity of our counting system. It was indeed so to its inventor, the Stone-Age man. In our times though, we are free to create a counting system of our own: limiting ourselves to the whole numbers, we can decide to start at any point and to have that counting system incremented by any number, not necessarily 1. This is what mathematicians call arithmetic, geometric and harmonic progressions. Since we will concern ourselves only with counting systems that in a way look like arithmetic progressions, there follows an example:

7, 11, 15, 19 ...

For the purpose of the present discussion, we will call 7 the *starting point* and 4 the *unity* as it is a system that takes 4 to get from *base* to *target*. Generally speaking, an arithmetic progression is a sequence of numbers such that the difference between the consecutive terms is constant. In the example above the constant is 4, which is identical to what we have called *unity*.

Going a step further, one can decide to create a counting system that has a *variable unity*, not a fixed one. An instance of such a counting system could be:

9, 12, 17, 20, 25, 28, 33, ...

Unity varies here alternatively between 3 and 5, starting with 3:

9  + **3** = 12,

12 + **5** = 17,

17 + **3** = 20,

20 + **5** = 25 ...

For the sake of clarity, this counting system is said to have 9 as the *starting point* and 4[3,5] as *unity*. It is worth noticing that in the latter notation "4" stands for the average and that "[3,5]" stands for the increments that the counting system knows as well as for the order in which the increments are taken, in this case starting with "3", then "5". Finally, it should be noticed that a sequence of numbers that has a variable unity does not comply with the given definition of arithmetic progression as the difference between the consecutive terms is not constant.

Coming back for a while to what has been called a sequence of numbers with a variable unity: first of all, the very concept of "variable unity" is unknown in mathematics, and secondly, such a sequence of numbers, even if it has a harmonious build up, is also unknown in mathematics. Since things that do not exist have no name, it seems plausible to provide such a sequence of numbers with the general name of *counting system*. After all, any sequence of numbers, even the most erratic one, can be granted such a name provided it is used to count something.

The previous considerations were intended to open the way for the discussion concerning the essence of prime numbers. These have always been defined as those numbers of our counting system that are divisible only by themselves and by unity. The mathematicians' unity, being the number 1, has always been excluded from the family of primes; whereas the number 2 has been admitted, and it is the only even number considered to be prime.

Looking at a sheet of paper with as many consecutive numbers as one can write on it and wondering about the erratic distribution of the prime numbers in our counting system is fascinating. Their distribution seems to obey no law. The lawlessness of their occurrence has undoubtedly to do with the intrinsic imbalance of the Stone-Age counting system which, if provided with the number zero, is identical to that of current science. This imbalance has to do, as has been explained, with the chosen unity, the number 1.

There is, however, an alternative way of defining prime numbers: **prime numbers are numbers that do not fit in a proportionally ordered counting system, i.e., in a counting system with variable unity.** Now arises the problem of deciding what a proportionally ordered counting system is. Well, it is a counting system with variable unity.

[*The reader has undoubtedly noticed that I have been using mathematical concepts that are unknown in mathematics. One of those concepts is that of a **counting system**. To my knowledge mathematicians know just one counting system, namely theirs: 0, 1, 2, 3... inherited from Stone-Age man. Since I am using sequences of numbers that are definitely not progressions, but that in a way look very much like the scientists' counting system, I have decided to give them the same name.*

*An identical reasoning has been applied to the concept of **unity**. Science knows just one unity: the number 1, a constant. While in search of an efficient way to find prime numbers, I came across sequences of numbers that had a double, intermittent increment. As the number 1, the classical unity, has no other function* *than that of being an increment, I decided to call also **unities** the increments of those numerical sequences. Since there were more than one, I decided to call them a **variable unity**.*

*Having arrived at this point, I have to confess that for many years, I was convinced that the apparently random distribution of prime numbers within the scientists' counting system was due to the imbalance that the scientists' unity (the number 1) brings to that system. I therefore intuitively concluded that the key to solve that problem would be to find a counting system based on a unity other than the number 1.*

*After many trials, and indeed, more errors, I reached the conclusion that no constant unity would do the job, and that is where I conceived the idea of a variable unity. Once I had reached this point, it was a matter of minutes before intuition led me to conclude that a counting system with the unity 3[2, 4] and starting point 5 was the sequence I was looking for. Such a sequence looks like this:*

5  7  11  13  17  19  23  25  29  31  35  37 41  43  47  49   53  55 ...

*It is actually a counting system equal to that of the scientists', stripped of "their" number zero, "their" unity (the number 1), and of all multiples of the numbers 2 and 3. I decided to call it "the basic sequence".*

*Again, led by intuition, I had reached the conclusion that prime numbers were all nested within this basic sequence, not to be "computed", but to be "found". In other words, in the counting system I was looking for, primes and non-primes were to be found by the **rank** they occupied in that system, not by the **amount** they represented. In other words, numbers in the basic sequence were not **cardinal**, but **ordinal**. So, whether, for example, 997 is prime or non-prime should be determined by the place it occupies within the basic sequence, not by being the number 997, three units away from 1000. In a sequence of numbers with variable unity, those numbers partially lose their cardinal identity and become ordinal, also partially.]*

The basic sequence that has just been spoken of, having a variable unity with an average of 3, has the following property: applying a so-called integer division by 3 to any one of its members delivers its ordinal place in that sequence. (As it often happens, "\" shall be used here as an operator for integer division. Applying "integer division" to a whole number means that after dividing that number by another number, one disregards not only the decimal point but also whatever decimal quantity the division has rendered, if any). For instance, 7\5 =1. So integer division by 3 of the *basic sequence*'s members delivers:

5\3 = 1  meaning that 5 is the first number of the

basic sequence.

7\3 = 2 meaning that 7 is the second number of the basic sequence.

11\3= 3 meaning that 11 is the third number of the basic sequence.

---

43\3= 14 meaning that 43 is the 14th number of the basic sequence.

Actually then, as all that matters to the members of the *basic sequence* is their rank within that sequence, one could transform it into a sequence that looks like this (with the corresponding members of the *basic sequence* underneath):

1  2  3  4  5  6  7  8  9  10  11  12  13  14  15

5  7  11  13  17  19  23  25  29  31  35  37  41  43  47 ...

Looking at the basic sequence it becomes obvious that:

-    by eliminating all the multiples of 2 between $2^2$ and $3^2$ all the numbers left over are primes.

-    by eliminating all the multiples of 2 and 3 between $3^2$ and $5^2$ all the numbers left over are primes.

-    by eliminating all the multiples of 2, 3 and 5 between $5^2$ and $7^2$ all the numbers left over are primes.

-    by eliminating all the multiples of 2, 3, 5 and 7 between $7^2$ and $11^2$ all the numbers left over are primes.

-    ...

This is actually a very old procedure for "discovering" prime numbers and is called the Sieve of Erastothenes. It is very well known that this method of finding primes becomes highly inefficient as numbers grow larger and larger. These inefficiencies are very well known so there is no reason to treat them here.

As stated before, the simplicity of the *basic sequence* as an instrument to find prime numbers resides in the fact that all non-primes occupy predictable places in that sequence and that by eliminating them, without having to make any computation whatsoever, the numbers that are left over are prime numbers. All primes will be there except those that are < 25. All one has to do is:

1.    Create a counting system with the unity 3[2, 4], with 5 as the starting point, and make it as long

as desired. Call this the **basic sequence**.

2.      Find the position of $5^2$ ($25\backslash3 = 8$) in that **basic sequence** and, starting at that $8^{th}$ position, check in the **basic sequence** all numbers that hold positions indicated by the counting system with the unity 5[3,7] and with the starting point $5^2$ (i.e., the $8^{th}$ position in the basic sequence). Checking numbers in the basic sequence could be done by converting them to 0, to 1, to a negative number … as long as the checked numbers are distinguishable from the unchecked ones.

3.    Repeat analogously with 7, 11, 13 … what was done with 5.

4.      At the end of this iterative process, the original basic sequence will contain both checked and unchecked numbers. The unchecked numbers are the primes. All of them.

Step 3, above, is NOT as simple as it looks. Looking back under point 2 at "unity 5[3,7]", 5 is said to be the base (b), 3 is said to be the "small step" (ss) and 7 the "big step" (bs). So, "unity 5[3,7]" could be rewritten generally as: "unity b[ss,bs]". Well, as it appears, the checking process in the basic sequence obeys the following rule:

5[ss,bs]

7[bs,ss]

11[ss,bs]

13[bs,ss]

Etc., where:

ss = previous ss + 2 (except with 5, of course)

bs = 2b − ss

The power of this simple algorithm resides in the fact that once the basic sequence has been created, the only variable that needs to be computed is $b^2\backslash3$ each time a new checking round begins. The checking itself of non-primes is done based on the *position* they occupy in the basic sequence, not on the *cardinal number* they represent; so no computation whatsoever is required in order to find them. Of all existing algorithms aimed at computing prime numbers, the one that has just been described is by far the most efficient.

For computer programmers' sake, the prime numbers' generating algorithm will be repeated. Comprehension is partially left to intuition. This algorithm will generate all primes > 25, so the programmer should take care of generating all primes < 25.

1. *Generate a sequence of n zeroes, Z, addressable at bit level, where n can be as long as desired.*

2. *Find in Z position $(5^2 \backslash 3 = 8)$. Generate $S_1$, a counting system 5[3,7] with starting point 8 and convert to 1 all positions <= n in Z indicated by $S_1$.*

3. *Find in Z position $(7^2 \backslash 3 = 16)$. Generate $S_2$, a counting system 7[9,5] with starting point 16 and convert to 1 all positions <= n in Z indicated by $S_2$.*

4. *Find in Z position $(11^2 \backslash 3 = 40)$. Generate $S_3$, a counting system 11[7,15] with starting point 40 and convert to 1 all positions <= n in Z indicated by $S_3$.*

5. *Find in Z position $(13^2 \backslash 3 = 56)$. Generate $S_4$, a counting system 13[17,9] with starting point 56 and convert to 1 all positions <= n in Z indicated by $S_4$.*

   *...*

   *The sequence 5, 7, 11, 13 ... is nothing other than $S_0$, a counting system 3[2,4] with starting point 5.*

   *Once this converting process is finished you*

*are left with Z containing "zeroes" and "ones". The prime numbers are held by all those positions that still hold a zero. Call this position p. To find the corresponding prime number PN take PN = (p\*3)+2 if p is odd. If p is even, take PN = (p\*3)+1.*

Looking at both versions of this prime numbers' generating algorithm, one has to conclude that there are a whole series of counting systems of the type b[step1,step2] with a particular starting point, that play an essential role in *finding* the prime numbers' places, not in *computing* them. These numeric sequences are unknown in mathematics, with partial evidence found in the fact that they do not even have a name.

The distribution of prime numbers in our numerical system is determined by the *position* they occupy there. They are found, be it by exclusion, in perfectly predictable places. Looking at them in this way, one has to conclude that prime numbers have nothing to do with indivisibility of any sort but with being "out of order" in a Stone-Age numerical system that is still in pretty good shape and that will keep itself in good shape provided science keeps its hands off of it.

# REFLECTIONS

From what has been written in the first three chapters, there are many conclusions to be drown and many questions to be asked. By far the most important of all the questions is: how reliable is our counting system when it comes to using it to represent and to measure Nature?

Science has no doubt achieved remarkable results and will continue to do so as long as we continue to be the dominant species on Earth and its surroundings. We are great engineers. The problem of science though is multiple:

- It has used uncritically a counting system invented by our ancestors thousands of years ago that was aimed at measuring discrete variables. Science, however, has made use of the same system to measure continuous variables.

- Once continuous variables had come into existence, science introduced the number zero, placing it at a distance from 1 that was supposed to be equal to the distance between any two consecutive numbers.

- The extraordinary engineering achievements obtained by using such a continuous counting system with the number zero attached to it led science to believe that Nature's fundamental variables were ordered in accordance to our Stone-Age counting system as it had been enhanced by science, and to believe that we therefore were entitled to measure Nature with it.

The greatest fallacies of this way of thinking are:

- Nature does not admit a zero measurement point in its fundamental variables. While measuring, numbers are supposed to be marking points on any scale used to measure variables, be they discrete or be they continuous. Well, at "zero point" any variable has long ceased to exist, except for engineering purposes where any point, mainly in the time/space continuum, can arbitrarily be marked with the number zero.

- While measuring the fundamental variables of Nature, *equidistant* between the number zero and the number 1 on the one hand, and between any two other consecutive numbers of the counting system on the other, is false; for it leads to Nature's *intermittency*.

- Zero is as far from the smallest number we can imagine as "infinite" from the largest possible number we can conceive. So zero is as unconceivable to the human mind as "infinite" is. They are both unreachable to our brain.

- Accepting zero as a measurement point in any fundamental variable of Nature is incompatible with its *measurability*, measurability being nothing other than the repeatability of the chosen measurement unity on the ratio measurement scale. Departing from the zero point, one never ever reaches point 1 no matter how tiny the measurement unit is, for reaching such a point leads necessarily to the possibility of reaching measurement point 2, and this reduces any variable to intermittency, as we have seen.

Finally the big question arises: what then is science's number zero? In these pages it has been suggested that such a point is a sort of convergence of all variables, both discrete and continuous. At zero energy, there is not only an absolute absence of any energy, but also of anything else: it is an absolute vacuum. The same holds for zero time, zero space or zero goats for that matter. It is not even the departing point of any of those variables, it is the absolute absence of anything at all.

That zero is an absolute vacuum becomes clear considering this: suppose for a moment that zero mass contains indeed no mass at all, without being an absolute vacuum. If the latter is the case, then zero mass should contain at least the tiniest possible remains of any other fundamental variable of Nature, be it time, be it energy ... In this case, the fundamental step from zero mass to mass does NOT take place as mass does NOT originate from zero mass but from time, from energy ... from the tiniest possible remains of any variable that was present at zero mass, at least if zero mass was not equal to absolute vacuum.

This chapter began with the question: how reliable is our counting system when it comes to using it to represent and to measure Nature? Well, we must say that although it has a great reliability for engineering purposes, it is completely unreliable when it comes to using it to measure the fundamental variables of Nature.

And the reason is: zero, as a number, is hidden from us. Measuring time, space, energy … has necessarily to begin *after* such variables have come into existence, once they have left their zero measurement point behind. Only from then on are we allowed to begin measuring them with our Stone Age counting system enhanced with infinitesimal steps between any two consecutive marking points. In a few words, deprived of their zero value, Nature's continuous variables are measurable by the human mind only at *interval* level. This should surprise no one considering for instance what happens to temperature: as stated earlier, we have apparently been able to find this variable's lowest possible value at -273.15 degrees Celsius. With this Nature-given jewel in hand, why are we still unable to establish a measurement unity that enables us to make measurements at ratio level on temperature? And then, extrapolating to time: suppose one day Nature reveals to us zero time. Our clocks then start running. How can we be sure that the time elapsed between 00:00:00 and 00:00:02 is twice as long as the interval that runs between 00:00:00 and 00:00:01? And then, supposing a 2-second interval is twice as long as a 1-second interval: where does this certainty come from? Is it because time reveals itself to us better than temperature does, or is it because our brain is better equipped to understand time than it is to understand temperature? Why are we not able to find in temperature a unity that makes it fit to be measured at ratio level whereas with time *any* chosen time-length will do?

Science has been claiming for the past centuries to be able to measure Nature from the zero point on, including the number zero itself. What is then this zero measurement point science has claimed to have found?

It should be noticed that science's number zero has the following properties:

- It is supposed to exist.

- Any variable's zero measurement point is supposed to be different from any other variable's zero measurement point.

- There is continuity between the zero measurement point and the least possible, measurable value of the variable being measured.

- The gap between the variable's non-existing point (its zero measurement point) and the least possible measurable value of the variable concerned occurs only once, namely right after the variable has left its zero measurement point. Nonetheless the variable is measurable, and any measurement unity that is chosen does not lead to intermittency.

So, we have to deal with a zero point:

- that does exist.

- that originates all fundamental variables of Nature.

- that diversifies all those fundamental variables of Nature at the zero measurement point, i.e., even before they become measurable to the human mind.

- that gives origin once and for all to all fundamental variables of Nature making them measurable, and impeding intermittency.

In trying to identify the essence of science's number zero, there is only one solution: the Number Zero is God, the origin of everything, the only being able to diversify Nature before its fundamental variables come into existence, the only being capable of creating time, not from zero time but from absolute nothingness, capable of creating space, not from zero space but from absolute nothingness, capable of creating light, not from zero light but from absolute nothingness ... a being that is capable of discerning between the nothingness that gave birth to time, the nothingness that gave birth to space, the nothingness that gave birth to light ... But most important of all, this is not the God preached by any religion or by all of them for that matter: this is the God whose existence is necessarily postulated by the most fundamental laws of Nature ... as these have been conceived by science itself.

***

Converting science's Number Zero to God, the Origin of Nature, changes dramatically the essence of the fundamental, continuous variables of Nature. Deprived of their zero measurement point, time, space, energy, light ... become floating variables as it were, with no beginning and no end. This should surprise no one: measuring, as has been said time and again, is nothing other than attaching numbers to variables. Well, numbers are all terms of our numerical system, a system that certainly has no end, but a system too that has no beginning even though we have thought for centuries to have found this beginning at point zero.

Deprived of their "roots" in the number zero, what do these variables still represent? Knowing that under such circumstances, time, space, energy, light ... can be measured only *after* they have come into existence, i.e., after they have left their zero-status, at which point of their own continuum do these variables lend themselves to allow the human mind to start measuring them? But irrespective of the position of that primordial measurement point, one should never forget that we are then measuring light in *already* existing light, space in *already* existing space, time in *already* existing time … and besides, looking at the words *after* and *already*: why is it that we human beings are equipped to conceive of a variable's coming into existence only within the fundamental variable of time? Deprived of its zero origin, we can conceive space's coming into existence only in time, not in space itself, not in energy, not in mass. All this is due to our ability to conceive of *any* change as occurring *only* in time. A change of the fundamental variable of time that takes place in the fundamental variable of energy is as unconceivable to us as is a change of the fundamental variable of light that takes place in the fundamental variable of light itself, or in mass, or in gravity. All these changes, so we are told by our brains, can occur only in time.

So, looking at any fundamental step from zero energy to energy, from zero light to light, from zero gravity to gravity … our brain dictates that such a step can occur only in time. But generally speaking, *any* change that takes place in *any* fundamental variable can be conceived by human beings only as taking place in time. If a change occurs for instance in gravity, the only way for this change to occur "outside time" is to conceive time as a non-existing variable at the "moment" such a change takes place. But for us human beings, a change that takes place outside time is inconceivable. In such a case, we can only think of the situation "before" and the situation "after". On the other hand, if such a change takes place while time is already "running", the only way for such an event to take place "outside time" is that the situation prior to the change and the situation after the change occurred simultaneously, something our brain can't conceive; for in such a case, a particular situation should have to have existed and not existed at the same time.

What is so fundamentally wrong with our brain that we are bound to make time-dependent the very origin of all fundamental variables of Nature? Or is time the origin, not only of time itself but also of all fundamental variables of Nature? Or, on the contrary, is time not one of Nature's fundamental variables at all?

Be that as it may, we are equipped to know the fundamental variables of Nature as being distinct from one another: time is not energy, mass is not light, light is not time. We can of course *speculate* that they are all just one variable, that there are no fundamental variables at all, … but that is not the way our brain was made to know Nature. So unless Nature has been playing games with us, the fundamental variables of Nature *are* distinct from one another. Why Nature has forbidden us to *measure* those fundamental variables outside time, we will probably never know.

\*\*\*

After all these considerations, what happens to

the scales of measurement? If we stick to the four measurement levels explained in chapter 2, then we are bound to conclude that while trying to "measure" Nature's fundamental variables, we can't go further than the *interval* level as those variables, deprived as they are of their zero value point, do not admit measurement at ratio level.

At interval level though, equidistance is still needed and this requires a constant unity. The problem with unities is that although we are free to choose them, we have no other choice than to fix their size based on the knowledge we have at present of the fundamental variables of Nature, ignoring what might have happened to them right after they had left their zero-status if indeed they ever had such a status. To science, this is the more ominous as it pretends to be capable of conceiving those variables *at* their zero measurement point. Be that as it may, should the fundamental variables of Nature ever have come into existence, it could well be that those variables, undergoing the abysmal changes that then were taking place, were unfit to be measured by the unities we conceive now. How long did time take to become time? Did gravity come from absolute heaviness or indeed, from zero weight? Did light originate from absolute brilliantness or on the contrary, from absolute darkness? Did all these dramatic changes take place in time or was time not yet "there"? What does a second or a minute measure while time is still becoming time?

All in all then, discarding Nature's measurability at ratio level, the best we can do, as it has just been said, is to measure the fundamental variables of Nature at *interval* level with all sorts of unbelievable consequences, the most important of all being: the impossibility of carrying out ratio operations, not only within, but also between them. If time cannot be multiplied by time, if light speed cannot be squared, if mass is not divisible by energy ... What happens for instance to the mother of all equations:

$$E = mc^2 ?$$

One cannot discard the possibility that one day an extremely clever engineer will show that taking a discrete amount of mass and producing an amount of energy that will be equal to that mass multiplied by the squared speed of light. Whether the resulting energy will express itself  in temperature, in light brightness, in sound, in gravity ... it really does not matter. When that occurs, one should bear in mind that the taken matter did not originate at matter's absolute nil measurement point, and that the resulting energy gave origin to energy in Nature as both matter and energy had long before, in time,  come into existence.

So, coming back for a moment to Mr. Einstein's equation, one has to conclude that if there was mass to be taken, not as a constant but as a fundamental variable of Nature, multiplying that mass by the squared speed of light would lead not only to an incommensurable amount of Energy but also to contradictorily exclusive amounts of Energy, should Nature decide to repeatedly "re-create" Energy from Mass. How great will those amounts of Energy be, we simply are not equipped to know. Perhaps Nature is. After all: what does 12 degrees Celsius multiplied by 7 degrees Celsius yield? Any answer will do as no answer can be proved false AND no answer will do as any answer can be proven false. We live in an uncertain universe no matter how much we would prefer to dwell in a certain, predictable one.

Having arrived at this stage, is it not strange that, as chapter 3 suggests, prime numbers leave their mysterious, hidden places once they are placed in a numerical sequence that has been arranged, not in an *interval*, (let alone in a *ratio*) but in an *ordinal* way? Rather than being numbers that are divisible only by themselves and unity, are prime numbers not those that do not fit in a particular sequence that has been arranged in an *ordinal* way? Is *indivisibility* the most essential characteristic of prime numbers, or is it *ranking*? Are primes "foreign" numbers in a *ratio* world or in an *ordinal* one? Why are prime numbers blindly localisable in a numerical sequence that does not even meet the simple rules that mathematicians impose to their arithmetic progressions?

The lack of a zero point measurement of the fundamental variables of Nature, living in a universe where the highest measurement level accessible to human beings does not go beyond the interval level, a human brain that is unable to understand Nature outside time, fundamental variables of Nature that we are not allowed either to multiply or divide by themselves or by one another, prime numbers that leave their hidden places if searched for in an ordinal sequence AND if searched with ordinal sequences, simple, uncomplicated numerical sequences provided with terms separated from one another by a double unity ... If all this is true we shall have to ask ourselves whether the big book of Nature is written in the same language used by our engineers that are conquering the Earth and its surroundings. Most probably though, no matter how hard we try, our brains will never ever grasp the nature of Nature.

# The nature of Nature

# THE AUTHOR

German Navarro has authored several novels in Dutch, his own language. Besides, he has written two novels in Spanish and a book on spirituality, published in Italian. The present work is his first publication in English. German Navarro is Dutch, has lived and worked in several countries, one of them being Italy. Rome in particular, the former capital of the world, has been his habitat for the past decade or thereabouts.

In *The nature of Nature* the author invites the reader to reflect on some aspects of the fundamental variables that constitute our universe: time, mass, energy, light ... Are they what they seem to be? But most importantly, are they what science tells us they are? If the author is right, there is something fundamentally wrong both with the way science, including scientists like Mr. Einstein himself, claim to be able to measure our universe AND with the capability of our brain to understand the very nature of Nature.

To the reader interested in *prime numbers*, the author offers in this book an original solution to the problem. Primes cease to be hidden, mysterious numbers, randomly distributed in the counting system as if were, and become localisable at fixed, predictable places within that numerical system. The new algorithm described in this work leads to the fastest possible solution to that old problem of prime numbers.